AESTHETIC SURGERY

Also by M. Clement Hall

Non-Fiction

The Locomotor System — Functional Anatomy

The Locomotor System —Functional Histology

Architecture of Bone; Luschka's Joint; Lessons in Histology

Palestine —The Price of Freedom; Intifada

IME – The Word Book; Independent Medical Examinations

The Fibromyalgia Controversy; Washing Away of Wrongs

A Calendar of Miseries; Murder of Richard Hunne;

Modern Eye Surgery

Charles River Editions e-books

History of Afghanistan; History of Iran; History of Syria

Memoir

Viet Nam 1963; Viet Nam 64-66; Vale Viet Nam

Fiction

Trauma Surgeon; The Spare Parts Box; The King George Inn

Martin's Absolution; Martin in Byzantium

Farmer George; Diamonds in West Africa

AESTHETIC SURGERY

M Clement Hall

AESTHETIC SURGERY

Copyright © 2012 by M. Clement Hall.

All rights reserved.

No part of this publication may be reproduced, stored in a retrieval system, or transmitted in any form or by any means, digital, electronic, mechanical, photocopying, recording, or otherwise, or conveyed via the Internet or a Web site without prior written permission of the author, except in the case of brief quotations embodied in critical articles and reviews.

ISBN 978-1-105-90230-7

CONTENTS

Anatomy And Terminology Of The Breast………...………7

Breast Augmentation………………………………….……9

Breast Lift (Mastopexy)……………………………….…..11

Breast Reduction, Reduction Mammoplasty……………...13

Abdominoplasty…………………………………………..15

Face Lift, Anatomy Of The Face…………………………17

Face Lift, The Aging Face………………………….…....21

Facelift, The Procedures…………………………………..23

Liposuction…………………………………………….…..25

Botox…………………………………………………...…29

Nose Surgery, Rhinoplasty………………………………...31

Non-Surgical Rhinoplasty………………………………....35

Revision Rhinoplasty……………………………………..35

Eyelid Surgery, Blepharoplasty……………………………37

Cervicoplasty, Neck Lift……………………………….....39

ANATOMY and TERMINOLOGY of the BREAST

The protruding portion of the breast varies in shape, position and size, but the position and extent of the base of the breast remains fixed throughout life. The chest (thorax) has 12 ribs on each side, the base of the breast extends from the 2nd to the 6th ribs, centered on a line drawn down from the middle of the collar bone (mid-clavicular line).

At the base of the breast there is a layer of relatively loose superficial fascia, then deep to that is the submammary (retromammary) space where the draining lymphatic ducts are found. Under them is the firm layer of deep fascia which envelops the chest muscles (pectoralis major and minor).

The breast (mammary gland) begins its life as the same structure in both males and females, in the male its development generally regresses, in the female it advances, so the initial single duct develops into a structure of between 15 and 25 lobes, each with a *lactiferous duct* leading to the central nipple, and with a dilatation (*sinus lactiferous*) beneath the areola, the pigmented disc of the skin surrounding the nipple.

Loose connective tissue lies between the lobes, and the whole sphere is enveloped in a layer of fatty subcutaneous tissue which obscures the shape of the individual lobes.

The breast is supported in position by so-called ligaments; whereas most ligaments in the body run from bone to bone to stabilise a joint such as the knee, in the breast these bands of fibrous (*collagenous*) tissue run from the deep underlying fascia to the skin (*dermis*). They were described by an anatomist-surgeon Astley Cooper, whose name is given to them. As the breast becomes heavier with fat, or in lactation, the ligaments are stretched, and stretched ligaments of this nature have no mechanism for regaining their prior length, so once lengthened they remain lengthened and the breast loses its initial erect position, a condition known to the vulgar medical student as "Cooper's droop," or in better terms as "mammary ptosis." The degree of ptosis, or sagging, has been defined for convenience as

Grade I Mild: nipple is located below the inframammary fold (IMF), but is above the lower pole of the breast.

Grade II Moderate: nipple is located below the IMF, not completely below the lower pole of the breast.

Grade III Severe: nipple is far below the inframammary fold; no breast tissue below the nipple.

BREAST AUGMENTATION

Why Augment?

This is a procedure distinct from "reconstruction" after injury or removal of a breast. It is usually performed when the patient is dissatisfied with the size of her breasts, may have been teased at school, or suffered negative comments from male companions, or may think she would "present" better in the job market if she had a better figure. As in all forms of aesthetic surgery, the surgeon must be aware of the patient's expectations and psyche.

What with?

There are standardized implants approved by the FDA for use in the USA, but since doctors in North America frequently are called on to treat persons who have had surgery in other countries it is as well to know that "elsewhere" implants may be made of many types of material including balls of ivory, rubber, animal cartilage and various types of sponges.

Saline implants: the container is a silicone elastomer (meaning "rubber-like") that is filled with salt water either prior to or after implantation.

Silicone gel implants: the container is a rubberized silicone, but the contents are in place prior to implantation. The device has gone through a number of stages (*generations*), each representing an improvement or a correction of a perceived deficit, the worst of which were leakage of silicone and fear of cancer.

How?
Since this procedure is performed to enhance appearance, it is highly desirable that the incision should be of minimal size and the scar should not be not noticeable, bearing in mind that those with darker skins are more likely to have visible scars. The prosthesis that is intact prior to insertion (polymer gel) will of necessity require a larger incision than the prosthesis filled after insertion (saline).

Standard incisions (surgeons are imaginative and many variations are possible):

> *Periareolar:* a two inch incision made around the inner aspect of the areola
>
> *Inframammary (IMF)*: made under the breast at the infra-mammary fold, hidden by the slightly pendulous breast, but large enough to insert the pre-filled silicone-gel prosthesis.
>
> *Remote tunnelling*: these may be from any direction the surgeon chooses, umbilicus, armpit etc., and are performed either bluntly or with the endoscope; the scar is remote, the surgical control is diminished.

Placement of the prosthesis:

> *Subglandular*: the implant is inserted into the submammary space deep to the breast and superficial to the fascia over the muscle.
>
> *Subfascial*: inserted between the deep fascia and the pectoralis major muscle.
>
> *Submuscular*: inserted deep to the pectoralis major muscle, whose lower attachment to the ribs may or may not be partially severed.

BREAST LIFT (MASTOPEXY)

Definition

This procedure is to replace, possibly reshape the breast into a more desirable position. Although it does not by definition include prosthetic breast enlargement, that may be performed at the same time.

Who?

Anyone who is not satisfied with the position of either or both breasts is a candidate, but typically the operation is performed on a woman who was previously satisfied with the position of her breasts, but the Cooper's suspensory ligaments have lengthened following pregnancy, weight gain then loss, or aging, and the nipples are lower than she (or her partner) would wish. The contents of the breast may not be interfered with by the surgery, and lactation in a subsequent pregnancy will remain possible.

How?

Periareolar lift: a crescent of skin is removed above the areola.
Lollipop lift: incision is around the areola and vertically down to the lower margin of the breast.
Anchor lift: there are three incisions which collectively resemble the shape of an anchor. They are taken around the areola, vertically down from the lower margin of the areola, and across in the in the inframammary fold.

Combined with Augmentation Mammoplasty
An implant procedure might be sufficient to compensate for Grade 1 ptosis, if considered insufficient, the periareolar incision is appropriate for a combination of both implant and lift procedures.

Potential Complications
There are the problems associated with any form of surgery, incisions and anesthesia, and there are specific issues of potential necrosis (death) of the nipple and areola, changes in perceived sensation in the breast, and asymmetry (difference in size) of the two breasts.

BREAST REDUCTION
REDUCTION MAMMOPLASTY

Who?

Overly large breasts have appropriately been described as an "affliction."

Terms have been derived, ranging from macromastia to gigantomastia which may not be etymologically correct but are certainly descriptive.

Reasons for wanting a reduction:

Aesthetic and psychological, for instance asymmetry, unable to wear wished for clothes, butt of jokes, self-image and self-esteem.

Due to the weight of the breasts pulling the bra straps deep into the shoulders, provoking postural change and backache, reducing pleasure in physical activities, interfering with breathing.

Hygiene: despite every effort to control moisture, skin beneath the breast becomes chronically inflamed and infected (intertrigo).

How?

The technique employed will depend on the situation, the amount of reduction required, the preferences of the patient and the decisions of the surgeon. To the extent possible, whatever the skin incision, the nipples and areolas will be left attached to underlying breast tissue, thereby retaining normal sensation, and the potential for breastfeeding.

Incisions are much as described for breast lift:

Anchor incision: around the areola, vertically down to the inframammary fold and transversely under the breast.

Lollipop incision: around the areola and vertically down to the lowest margin of the breast.

Horizontal incision: along the inframammary fold.

Once inside the breast it is a matter of skilled judgement to determine the amount of tissue to be removed, and at the same time not interfere with the areola, nipple complex (NAC), which is replaced in a higher position on the reconstructed breast, and then do the same to both breasts. Excess skin is cut away when the incisions are sutured.

Other variations in surgery are:

Free nipple-graft technique: potentially allows greater ease of access for removal of a larger quantity of breast tissue, but the NAC will be rendered permanently insensitive.

Liposuction: appropriate when the breast is fatty, and not much needs to be removed; the incisions are small but so is the volume removed.

Complications

There are all the potential complications attending any surgery and anesthetic. Specifically there is a hazard to the viability of the NAC, and the difficulty in achieving perfect symmetry – for the latter, revision is possible, for the former, the dice are cast.

ABDOMINOPLASTY

Abdominoplasty sounds medical and "tummy tuck" sounds friendly – both refer to an ill-defined operative procedure in elective cosmetic (aesthetic) surgery, and the extent of the operation may vary from a small skin operation to a reconstruction of the wall of the abdomen. The indications are a patient who is dissatisfied with the appearance of the abdomen, but typically someone whose abdominal wall is now less shapely than it used to be, perhaps following weight reduction, deliberately or after pregnancy, perhaps associated with the aging processes.

Mini Abdominoplasty

Better termed "partial" because it is not "mini" in extent. The incision is made low across the abdomen where it will be hidden by a "bikini-sized" undergarment. The subcutaneous tissues are undermined and separated from the attachment to the deep fascial wall of the anterior abdominal musculature. Tension is applied to determine the degree of excess skin which is then excised from the upper margin of the wound. The umbilicus (belly button) may be separated from its surrounds and reinserted (either as a free or pedicled graft) into a more normal anatomic position. A minor degree of tautening of the abdominal fascia might be undertaken. Some lipectomy by liposuction might be employed as indicated to improve contouring.

Full Abdominoplasty

The incision is in the same suprapubic location, but extends further to each side in order that a wider degree of exposure can be obtained after the subcutaneous tissue is separated from the deep muscle fascia of a larger proportion of the anterior abdominal wall. The fascia is plicated and sutured into a more taut position, possibly in a vertical line, possibly also by horizontal plication. The umbilicus is replaced. Liposuction is used as required. The excess skin is excised.

The procedure might be extended further to each side, to facilitate similar liposuction and excision of surplus skin over the lateral (outer) aspects of the thighs.

Add-ons

The other procedures performed under the same anesthetic are a matter for patient wishes and surgical judgement. If there has been very severe weight loss it is probable the patient will also wish to have a "buttock lift," which might be performed through extensions of the same incision. Or the patient might wish to have multiple separate procedures performed under the same anesthetic, popularised on TV as a *mommy makeover.*

Risks

There is a standard joke in surgery, which isn't a joke, "minor surgery is what other people have." Things go wrong. Things go wrong when you least expect them. There are hazards to anesthesia and to all forms of surgery. Your surgeon will explain these to you, and you will be expected to acknowledge that you understand them.

FACE LIFT
ANATOMY OF THE FACE

Skeleton

The front of the skull, the bones that underlie the soft tissues of the face, are named from above down:

Frontal: forehead
Bones forming the interior of the orbit
Nasal bone and cartilage
Maxilla: upper jaw bearing teeth
Zygoma: cheek bone
Mandible: lower jaw bearing teeth

Muscles

Superficial muscle aponeurotic system (SMAS) is an appropriate term in use by reconstructive surgeons who operate on the face, but not one that as yet appears in anatomy texts. The facial muscles are in many ways individual units with individual functions, but more fundamentally they develop from a single sheet of muscle, are interconnected by relatively dense fascial (collagenous fibers) tissue, and can in reconstructive surgery be thought of as an intact unit when it comes to subcutaneous tautening of the facial tissues (distinguish between facial and the totally different fascial).

Muscles of Facial Expression, (grouped around the orifices of the face):

Eyes

Orbicularis oculi: a circle of muscle structured to close the eye as it contracts (a sphincter) formed in 2 parts; the *orbital* portion extends around the bony orbital margins of the face, and the *palpebral* part which attaches to the deep fascia of the upper eyelid.

Levator palpabrae superioris: functions in the opposite direction to the palpabral portion of orbicularis oculi, lifts the eyelid, and opens the eye; its muscle fibers run between the bone inside the upper orbital margin and down to the tarsal plate which gives substance to the upper eyelid.

Occipito-frontalis: this is an unusual double muscle, with its aponeurosis (flat tendon) between two muscle groups, one at the back of the skull (*occipitalis*) the other at the forehead – the *frontalis* portion. With this muscle everyone can wrinkle the brows, and some can wriggle their scalps!

Corrugator superciliaris: takes origin from the frontal bone at the upper rim of the orbit, runs through orbicularis oculi and frontalis and inserts into the skin near the inner end of the eyebrow; when contracted the eyebrows are pulled inwards and down producing the crease in the center of the forehead.

Procerus: a central muscle which arises from the nasal bone, passes up between the paired frontalis muscles to function as a depressor of the central brow and to create the horizontal creases at the root of the nose.

Nostrils

Compressor naris and *dilator naris* are sphincter muscles to open and close the nostrils (partially only in humans).

Levator alae nasi raises the wings of the nostrils, as in forced breathing, or picking up a scent. They are lowered effectively by elastic recoil, though there is a muscle whose nominal function is to do that.

Mouth

Buccinator is the muscle of the cheek; its fibers arise from the upper and lower jawbones, and pass transversely forward to the mouth; its function is to aid in chewing, moving the food around inside the mouth, but it is mentioned here because of its relationship with orbicularis oris.

Orbicularis oris is the sphincter around the mouth, and is a compound muscle with muscle fibers attached to upper and lower jaws, and others from the buccinators muscle. At the corner of the mouth (*modiolus*) the middle part of the buccinators fibers cross (form a *chiasma*) so upper middle fibers go to the lower portion of the sphincter and vice versa. When the muscle contracts the lips pucker into a "whistling" position.

Dilator muscles are in a sheet of fibers whose function is opposite to orbicularis, that is, they pull the mouth open and with practice can be made to contort the mouth or the corner of the mouth, in every conceivable direction.

Muscles of Mastication

These must be mentioned for completeness, but are not of consequence in "face lift" surgery. They function to move the teeth of the mobile lower jaw against those of the fixed upper jaw; *temporalis* can be felt at the temple, and *masseter* at the side of the clenched jaw.

Nerve supply

The facial nerve (VIIth cranial nerve) passes through the parotid gland to supply the muscles of the face. It is liable to be damaged in operations on the parotid which is not uncommonly a seat for cancer. Stroke (cerebro-vascular accident) is particularly likely to result in a weakness of one side of the face, which is often the first sign of its onset.

FACE LIFT — THE AGING FACE
WRINKLES & BAGS

There are a number of factors which collectively are responsible for the change of a face as it ages; underlying them is an actual change in the skeleton. In the maxilla which decreases in size as the skeleton ages, the orbits will increase, and the cheek fat pads and skin are displaced downward, resulting in deepening of the nasolabial fold. With loss of teeth from the lower jaw there is resorption of bone, the shape of the chin alters. Aging skin loses its elasticity and gravity plays a part as does reduction of muscle tone and specifically in the skin, changes due to years of overexposure to sunlight.

Known to the lay public as "wrinkles," or "smile lines," the surgeon gives them a name derived from the Greek – *rhytids*. No one uses this, but it comes into play with compounded technical expressions defining the changes that are to be corrected by surgery.

Not part of a face lift operation, but significant in considering incisions, particularly in the male, is the receding hairline which most persons of both sexes will experience.

Forehead
> this will become transversely wrinkled (rhytids) and drooped (*ptosis*)
> between the eyebrows *(glabella)* there develops vertical furrows (*glabellar rhytidosis*)

above the eyes droops (*brow ptosis*)

at temples wrinkles and droops (*temple rhytidosis and ptosis*)

Eyes

upper eyelid skin thickens and lid droops (*ptosis*)

lower lid redundant fat tissue (*pseudoherniation*) and wrinkles (*rhytidosis*)

"smile lines" at corners (*lateral canthal rhytidosis*)

Nose

wrinkles at root (*nasal root rhytidosis*)

tip droops (*ptosis*)

line between nose and cheek (*nasolabial crease*) deepens

Cheek

pouches (*malar bags*) and wrinkles (*malar rhytidosis*)

cheek sagging with changes due to fat atrophy, resulting in *facial rhytidosis*

vertical crease in front of the ears (*preauricular rhytidosis*);

Mouth

vertical creases in upper lip (*perioral rhytidosis*)

upper lip flatter and wider, while vermillion is thinner

Lower Jaw

drooping of chin pad (*ptosis*)

drooping of skin and fat at borders (*jowls*)

FACELIFT — THE PROCEDURES

There is no "one procedure fits all," it's not like an appendectomy where a piece of the body is removed. Face lift, a rhytidectomy on a large scale, is a balance between minimising the surgical interference and maximising the alteration, a reconstruction of the facial skin. As such it is on every occasion an agreement between what the patient would like to have changed, and how the surgeon thinks this can best be effected. Fundamentally, excess skin is removed, remaining skin is made more taut and the underlying supportive tissues may also be tightened.

How

It is "cutting" surgery – an incision is made, either one longer incision, or more than one short incision. Many of the variations, and much of the drive to gain maximal cosmetic (aesthetic) benefit, lie in the minimising and placement of the incision(s).

The positioning of the incision might, at the surgeon's judgement, be different in men than in women, because of the facial hair.

In principle, the incision is made far back in the face, anterior (sometimes even posterior) to the ear, and the facial tissues are lifted from the deeper tissues by undermining them.

When the skin and subcutaneous tissues have been lifted as a unit, necessary tightening up (*plication*) of the muscular aponeurosis is performed as the surgeon considers indicated. The skin of the face is then drawn back and up to the incision and the newly defined excess quantity is excised.

The incision is closed with whichever technique the surgeon believes will cause least scar, yet effect a firm closure of the skin margins – in general a choice between sutures or staples in their multiplicity of commercially available forms.

The "lift" might be supplemented by other techniques of fat or prosthetic grafting (*liposculpture*), by an element of liposuction (known also as *lipoplasty*), or Botox injection, or combined with other reconstructive procedures, such as on the eyelids.

When

This is a question to be decided by the patient making use of the advice of the experienced surgeon.

Most surgeons are likely to advise early rather than late procedures, to operate while the tissues have the consistency of relatively young material, and at a time when a smaller procedure or perhaps several relatively small procedures can be performed in preference to one large one.

LIPOSUCTION

If the Latin word is converted to English, the result is "sucking out the fat," and this gives an unfortunate impression that it's easy to do. Bit like taking a lump of wet clay and making a pot. Perhaps it *is* easy to do, but it's not so easy to do well. It makes more sense to call the procedure *lipoplasty* because it's a reshaping of the body, *recontouring* some say, and the suction out of the fat is merely how it's achieved, the skill lies in knowing which fat, and how much.

There are major issue of safety. Because it can be done under minor anesthetic, it has leant itself to performance by persons with minimal training who offer their services at a commensurate fee.

Fat

In the live human, fat is in a fluid state, but it is contained in cells called lipocytes. The suction removes cells and the fat they contain. It must be understood only fluid can be sucked out, more dense tissue, such as "cellulite" does not lend itself to aspiration (sucking out).

How

The aspiration is performed by a metal tube (*cannula*) attached to a controllable vacuum source, it's very simple and not unlike the domestic vacuum cleaner – and in its simplicity lies the danger – it looks like anyone could do it!

Small incisions are made, just big enough to insert the tube, and depending on the surgeon's plan, usually several incisions. There are many variations in technique, they depend to a large extent on the surgeon's choice and experience, and on the volume of fat to be removed. In *tumescent liposuction*, the area to be aspirated is flooded with saline containing local anesthetic and adrenalin (*epinephrine*) which reduces bleeding, but both have the potential of undesirable side effects. In *ultrasound assisted liposuction (UAL)* the ultrasound will render the fat cells more easy to aspirate; there are also laser assisted techniques which may reduce the bleeding

Who

One could reasonably say the ideal candidate is the person who hardly needs the operation – a person with a small amount of fat more than the ideal. The poor candidate is a person with older tissues that will collapse when the supporting fat is removed, and although grossly overweight thinks it can all be removed with a vacuum cleaner just like the dust in an abandoned room. These are the persons the unskilled operators treat, and the ones whose unfortunate outcomes hit the newspapers. Most patients are somewhere in between, but the procedure is intended to reshape a little, to trim the bulges that have resisted dieting; it is not appropriate for a large quantity of generalized obesity which should have been corrected by dieting.

NOTES

BOTOX

The word *Botox* is derived from *botulin toxin*, and botulin is derived from the Latin word for a sausage. It all goes back to the discovery that decaying sausage meat was due to a bacterial infection, and this bacteria produced a poison (toxin) which paralyzed the nervous system and resulted in epidemics of death associated with diseased sausages.

Roll forward a couple of hundred years and the toxin is now refined, and used in specific quantities to paralyse specific nerves and treat specific medical conditions. It's just as poisonous and dangerous as it ever was, but when used with knowledge and caution it has proved very useful. Nevertheless it remains a matter of great concern for those who are involved with issues of bioterrorism.

The use of Botox is the most common of the procedures in aesthetic surgery, perhaps because no surgery is involved! It is directed at the removal of wrinkles (rhytids), in particular at the forehead and above the bridge of the nose, and several millions of treatments have been given in the United States, with minimal ill effects, but deaths have been reported after the use of Botox for medical purposes other than aesthetic.

The typical treatment takes about 10 minutes, is performed by several injections into the forehead to paralyse the muscles that cause the wrinkles; the duration of the effect is going to vary from person to person, but usually requires further injections somewhere between 3 and 6 months.

Following the paralysis of the forehead muscles, the brows are smooth, lose the ability to be used in facial expression. There is some interesting research that suggests the inability to express emotion may have a reverse effect and dampen the actual feeling of emotions. The completely smooth brow is also seen by some as unnatural and it is not difficult to guess which person in the public eye, for instance on TV, has been treated.

NOSE SURGERY— RHINOPLASTY

Among the first surgical procedures recorded in history are techniques for restoring a missing nose, a problem due not only to disease but also to the use of nasal amputation as a form of punishment. Although there are still occasions when plastic surgical procedures are required for the correction of cancer and other diseases, most operations are performed for cosmetic reasons, and reshaping the nose (*rhinoplasty*) is now rated as the overall second commonest procedure in facial plastic surgery; among men, and patients under the age of 25, it is the most frequently performed

Who

Rhinoplasty, a plastic surgical procedure performed on the nose, is in terms of this article performed for cosmetic purposes on the external nose – the portion on view to the public. The essential difference between rhinoplasty and rhitidoplasty (face lift) is the former alters the original whereas the latter attempts to restore the original. Perhaps more than any other form of cosmetic surgery, it is wise to ensure the patient has reasonable expectations of the result of the operation, and the patient does not attribute personal, social and business failures to the shape of the nose with an expectation that all will be reversed when the nose is a different shape.

The nose is the central feature of the face, and although there are ethnic variations considered perfectly normal in their own society, a patient might be extremely conscious that their nose is markedly different from society's concept of "ideal" and wish to have it altered. There are post-traumatic problems in a class of their own, but in general the indication is the patient's belief they would "feel better about themselves" if the nose were different, and they should know exactly what difference they seek. For many clinics, computer imaging software is available which gives the prospective patient an opportunity to see the changes that could be brought about and then have the ability to decide whether that is what they seek.

When

Reconstructive procedures for birth defects, injuries, disease etc., are performed when indicated by the condition. Cosmetic, elective, procedures are usually delayed until growth of the facial skeleton is complete, that is in girls about the age of 16, and in boys 18.

How

Essential to surgical success is pre-operative planning. The surgeon will have mentally, if not on computer or paper, have subdivided the external nose into its many constituent parts and will know before any surgery starts exactly how he intends to change (or not change) each of those portions.

The operation may be performed under local or general anesthetic, usually the tissues are infiltrated with adrenalin (epinephrine) to reduce bleeding, and if under local anesthetic will have been injected with that agent, or regional nerve blocks will have been performed. The skin and subcutaneous tissues are elevated from the underlying cartilage-bone framework, the framework is revised as indicated by removing or raising portions with graft, the soft tissues are trimmed as indicated, the wounds closed and the nose splinted to keep the tissues at rest for best possible scars with healing.

NOTES

NON-SURGICAL RHINOPLASTY

There is a "non-cutting," hence "non-surgical" procedure by which the shape of a nose can be altered, resculpted, but only in the sense of addition, not subtraction, but by addition around some prominences they can be rendered less noticeable. The material injected is a proprietary form of calcium hydroxyapatite, the mineral found in bone. The theory is that it will provide substance for natural collagen to grasp onto and the change will become permanent. The same material used to "fill" creases has to be repeated after a few months. It would however, at little discomfort, give the patient an opportunity to "live with the new nose" and determine if that's what they want before a more permanent procedure is undergone.

REVISION RHINOPLASTY

The apparent triviality of the *fifteen minute nose job* fails to show that all procedures carry risks, either complications from the procedure or dissatisfaction after it has been performed.

Secondary Revision Rhinoplasty

It has become accepted in cosmetic facial surgery that after the elapse of time an originally satisfactory rhinoplasty is no longer pleasing, and reconstruction is sought. It is estimated as many as one fifth of rhinoplasty operations performed will go on to revision, known also as a "secondary rhinoplasty." There are occasions when an injury has occurred, there are occasions when the patient has changed their mind about the most desirable appearance, and (of course) there are occasions when the original procedure might have been better performed. It is to be expected that current surgical techniques are an improvement on those performed decades ago, and one would hope for further improvement in the future – it requires a special type of person to believe we have achieved perfection.

The revision, secondary, rhinoplasty is the same as all revision procedures in surgery, substantially more difficult than the first time round. The surgeon will have to cope with scars (visible and hidden) from the primary operation, and there may be defects in tissue substance which will require grafting. Cartilage grafts, if soft, can be taken from the ear, or if hard, from the rib cartilages close to the breast bone (sternum).

EYELID SURGERY— BLEPHAROPLASTY

What

There are two "lids" to the eye, upper and lower; surgery on these has been called blepharoplasty since the procedures was first performed for the rare condition of cancer. It is now the most common of cosmetic facial operations.

The upper lid has a "skeleton" of a dense fascial plate and that usually is not at fault, nor the muscle that elevates it. With aging, however, the skin that covers the plate loses its turgor and if weight is gained there is increased fat deposited in the lid. The result is a general drooping of the lid which does not as a rule interfere with visual function but may be considered by the patient as unsightly, to be an early unwelcome evidence of aging, and to interfere with the application of eye make-up.

At the lower lid, it is less the lid that is involved than the skin beneath it which sags, "bags beneath the eyes" and is considered unsightly, although does not interfere with function.

How

The technique employed varies according to the situation, the patient's needs and the surgeon's preferences, but in general, it is performed under local anesthetic on an out-patient basis.

For the upper eyelid an incision is made along a skin crease, often an ellipse of skin is removed, the margin curving down medially and up laterally, and such excess fat as indicated is also taken out. Tautening of the underlying fascia and possible might be indicated, but as a rule is not.

For the lower lid, the incision may be made in a skin crease immediately beneath the eye's lower lashes if a portion of skin is to be removed to tighten up a "bag" under the eye. If it is planned to remove fat, and there is no need to remove skin, the incision might be made within the lower lid.

In both instances, lasers might be employed.

It is usual to combine blepharoplasty with other minor "touching up" operations, or sometimes it is the relatively minor accompaniment to a major total face lift.

CERVICOPLASTY — NECK LIFT

Aging changes in the neck

In essence, these are part and parcel of the changes in the face. The turgor of the skin diminishes with age and it hangs down under the influence of gravity; the effects of sunlight enhance the deterioration. The unusual muscle, *platysma*, found in the subcutaneous tissues of the lower face and neck (the horse uses his to shiver off flies) will also drag down the skin, and if present on both sides forms a framing margin for a central "turkey wattle." The platysma also drags down the tissues at the side of the jaw, causing "jowls," and, deep to the platysma, pads of "submental" fat develop under the jaw margin. As in the face, there are skeletal changes, in this instance the hyoid bone and the upper larnyngeal cartilages descend, with further loss of definition of the chin.

Who

The indications are the same as in a face lift, the operation(s) are performed when the patient is dissatisfied with appearance and when there are no contra-indications. It may be part of a more extensive "make-over."

How

A combination of procedures, or several separate procedures are possible, depending on the changes, the patient and the surgeon. Botox injections of the platysma might for a time serve the purpose. Liposuction might be all that is required if the issue is undesirable fat pads. If, however, there is significant sagging, then open surgery will be required – the "lift."

The procedure, variously called neck lift and cervicoplasty is also termed by some a *platysmaplasty* for that is the basis of the operation. Possibly under local, probably under general, anesthetic an incision is made beneath the jaw, another near the ear, excess fat is removed by liposuction, the subcutaneous platysma muscle is defined and tautened up to correct the droop, and the excess skin is excised as in a face lift.

www.ingramcontent.com/pod-product-compliance
Lightning Source LLC
Chambersburg PA
CBHW072305170526
45158CB00003BA/1196